Adding 1

$1 + 1 =$ ☐

$1 + 2 =$ ☐

$1 + 3 =$ ☐

$1 + 4 =$ ☐

$1 + 5 =$ ☐

$1 + 6 =$ ☐

$1 + 7 =$ ☐

$1 + 8 =$ ☐

$1 + 9 =$ ☐

$1 + 10 =$ ☐

Adding 2

$2 + 1 = \square$

$2 + 2 = \square$

$2 + 3 = \square$

$2 + 4 = \square$

$2 + 5 = \square$

$2 + 6 = \square$

$2 + 7 = \square$

$2 + 8 = \square$

$2 + 9 = \square$

$2 + 10 = \square$

PRACTICE PALS™

Write On Wipe Clean

Practice Book

FOR AGES 5 & UP

Addition & Subtraction

$1 + 2 = \square$

$1 + 5 = \square$

$3 - 1 = \square$

$6 - 1 = \square$

Makes learning fun by turning fun into learning!
Helps children learn the fundamentals of adding and subtracting numbers.
Educationally geared for pre-school through primary age groups.
Wipe clean surface allows kids to use again and again.

This book belongs to:

A message to parents:

The early learning years of a child's life are some of the most exciting and important he or she will ever face. They can also be a wonderful time of sharing and interaction between you and your child. We hope this Practice Pals™ Practice Book will serve as a part of that interaction. When your child first sits down to begin working in the practice book, here are a few things that you can do to create a positive learning experience:

- Have your child sit at a table in a comfortable position for writing.
- Use a non-toxic grease pencil, crayon, or dry erase marker for best results.
- Be sure he or she is holding the marker correctly for writing.

Most important of all, encourage your children. This book should be a recreation activity. Let them know it is okay (in fact, good) to make mistakes. Keeping it fun will keep them learning!

ISBN 1-930355-06-8
First published in the U.S. in 1996 by Greenbrier/Scentex.

Adding 3

3 + 1 = ☐

3 + 2 = ☐

3 + 3 = ☐

3 + 4 = ☐

3 + 5 = ☐

3 + 6 = ☐

3 + 7 = ☐

3 + 8 = ☐

3 + 9 = ☐

3 + 10 = ☐

Adding 4

$4 + 1 = \square$

$4 + 2 = \square$

$4 + 3 = \square$

$4 + 4 = \square$

$4 + 5 = \square$

$4 + 6 = \square$

$4 + 7 = \square$

$4 + 8 = \square$

$4 + 9 = \square$

$4 + 10 = \square$

Adding 5

$5 + 1 = \square$

$5 + 2 = \square$

$5 + 3 = \square$

$5 + 4 = \square$

$5 + 5 = \square$

$5 + 6 = \square$

$5 + 7 = \square$

$5 + 8 = \square$

$5 + 9 = \square$

$5 + 10 = \square$

Adding 6

$6 + 1 = \square$

$6 + 2 = \square$

$6 + 3 = \square$

$6 + 4 = \square$

$6 + 5 = \square$

$6 + 6 = \square$

$6 + 7 = \square$

$6 + 8 = \square$

$6 + 9 = \square$

$6 + 10 = \square$

Adding 7

7 + 1 = ☐

7 + 2 = ☐

7 + 3 = ☐

7 + 4 = ☐

7 + 5 = ☐

7 + 6 = ☐

7 + 7 = ☐

7 + 8 = ☐

7 + 9 = ☐

7 + 10 = ☐

Adding 8

8 + 1 =

8 + 2 =

8 + 3 =

8 + 4 =

8 + 5 =

8 + 6 =

8 + 7 =

8 + 8 =

8 + 9 =

8 + 10 =

Adding 9

$9 + 1 = \square$

$9 + 2 = \square$

$9 + 3 = \square$

$9 + 4 = \square$

$9 + 5 = \square$

$9 + 6 = \square$

$9 + 7 = \square$

$9 + 8 = \square$

$9 + 9 = \square$

$9 + 10 = \square$

Adding 10

10 + 1 = ☐

10 + 2 = ☐

10 + 3 = ☐

10 + 4 = ☐

10 + 5 = ☐

10 + 6 = ☐

10 + 7 = ☐

10 + 8 = ☐

10 + 9 = ☐

10 + 10 = ☐

Subtracting 1

1 − 1 =

2 − 1 =

3 − 1 =

4 − 1 =

5 − 1 =

6 − 1 =

7 − 1 =

8 − 1 =

9 − 1 =

10 − 1 =

Subtracting 2

$2 - 1 = \square$

$2 - 2 = \square$

$3 - 2 = \square$

$4 - 2 = \square$

$5 - 2 = \square$

$6 - 2 = \square$

$7 - 2 = \square$

$8 - 2 = \square$

$9 - 2 = \square$

$10 - 2 = \square$

Subtracting 3

$3 - 1 = \square$

$3 - 2 = \square$

$3 - 3 = \square$

$4 - 3 = \square$

$5 - 3 = \square$

$6 - 3 = \square$

$7 - 3 = \square$

$8 - 3 = \square$

$9 - 3 = \square$

$10 - 3 = \square$

Subtracting 4

4 − 1 = ☐

4 − 2 = ☐

4 − 3 = ☐

4 − 4 = ☐

5 − 4 = ☐

6 − 4 = ☐

7 − 4 = ☐

8 − 4 = ☐

9 − 4 = ☐

10 − 4 = ☐

Subtracting 5

$5 - 1 = \square$

$5 - 2 = \square$

$5 - 3 = \square$

$5 - 4 = \square$

$5 - 5 = \square$

$6 - 5 = \square$

$7 - 5 = \square$

$8 - 5 = \square$

$9 - 5 = \square$

$10 - 5 = \square$

Subtracting 6

$$6 - 1 = 5$$

$$6 - 2 = 4$$

$$6 - 3 = 3$$

$$6 - 4 = 2$$

$$6 - 5 = 1$$

$$6 - 6 = 0$$

$$7 - 6 = 1$$

$$8 - 6 = 2$$

$$9 - 6 = 3$$

$$10 - 6 = 4$$

Subtracting 7

$7 - 1 = 6$

$7 - 2 = 5$

$7 - 3 = 4$

$7 - 4 = 3$

$7 - 5 = 2$

$7 - 6 = 1$

$7 - 7 = 0$

$8 - 7 = 1$

$9 - 7 = 2$

$10 - 7 = 3$

Subtracting 8

$8 - 1 = 7$

$8 - 2 = 6$

$8 - 3 = 5$

$8 - 4 = 4$

$8 - 5 = 3$

$8 - 6 = 2$

$8 - 7 = 1$

$8 - 8 = 0$

$9 - 8 = 1$

$10 - 8 = 2$

Subtracting 9

$9 - 1 = 8$

$9 - 2 = 7$

$9 - 3 = 6$

$9 - 4 = 5$

$9 - 5 = 4$

$9 - 6 = 3$

$9 - 7 = 2$

$9 - 8 = 1$

$9 - 9 = 0$

$10 - 9 = 1$

Subtracting 10

$10 - 1 = 9$

$10 - 2 = 8$

$10 - 3 = 7$

$10 - 4 = 6$

$10 - 5 = 5$

$10 - 6 = 4$

$10 - 7 = 3$

$10 - 8 = 2$

$10 - 9 = 1$

$10 - 10 = 0$

Let's Practice!

Fill in the blanks to answer the problems.

$10 - 1 = \square$

$\square + \square = 17$

$\square - 6 = 2$

$8 + \square = 12$

$\square - 2 = 2$

$4 + \square = 10$

$10 + 3 = \square$

$7 - 3 = \square$

$7 - 4 = \square$

$5 - 2 = \square$

$\square - 2 = 1$

$2 + 2 = \square$

$1 + 2 = \square$

$\square + \square = 9$

$\square + 6 = \square$

$9 - \square = 5$

$\square - 5 = 2$

$1 + 6 = \square$

$\square + 4 = 11$

$4 + \square = 8$

$5 + \square = \square$

$2 + \square = 10$

$10 - 4 = \square$

$8 - \square = 1$

$\square + 8 = 14$

$3 + \square = 8$

$1 + \square = \square$

$8 + \square = 16$

$6 - 3 = \square$

$9 - \square = 8$

$3 + 9 = \square$

$1 + 9 = \square$

$8 - 3 = \square$

$4 + 6 = \square$

$5 + 10 = 15$

$7 + 5 = \square$

$12 - 4 = \square$

$\square - 1 = \square$

$10 - \square = \square$

$3 - 3 = \square$

PRACTICE POWER™

Practice Pals™ Practice Books are designed to introduce children to the basics of education by making learning fun. The colorful and fun activities of the Addition & Subtraction Practice Book help develop early math skills that are fundamental to all further learning experiences. The reusable write and wipe clean pages insure hours of fun, enjoyment and best of all... knowledge!

Look for these great fun and educational Practice Pals™ books!

... PRACTICE POWER™ educational products with Write On Wipe Clean format!

★ ...vel Teaser™ Boards ★ Lap Learner™ Boards

... Practice Pal™ Books ★ Flash Card Sets

Practice Pals™ Practice Books
Addition & Subtraction

© GREENBRIER/SCENTEX™
300 Greenbrier Road, Summersville, WV 26651
Made in USA

20004